oils & vinegars

oils & vinegars

Liz Franklin
photography by Richard Jung

RYLAND
PETERS
& SMALL
LONDON NEW YORK

Author's Acknowledgements

My thanks go out to the unsung heroes at RPS who helped to create a book of which I am immensely proud, and who continue to be a great bunch of people to work with, especially Alison, Steve and Céline. Thanks to Linda Tubby and Richard Jung for the mouth-watering photography.

Special thanks and oceans of love go out to the three most treasured men in my life: my wonderful sons Chris, Oli and Tim. Their constant encouragement with my writing, admirable efforts with a fork and willingness to suggest a tweak or two when needed are very precious to me. Huge squashy hugs too to my Mum and Dad for teaching me to delight in good, honest food and for their unswerving love and support. And to Laura, Paolo, Vicky, Junior, Max and Alessandro in Abruzzo, Italy, who make the most stunning extra virgin olive oil – and who are the most incredible friends a girl could ever wish to have.

Design, Photographic Art Direction and Prop Styling Steve Painter
Editor Céline Hughes
Production Manager Patricia Harrington
Art Director Leslie Harrington
Publishing Director Alison Starling

Food Stylist Linda Tubby
Indexer Hilary Bird

Note

- Eggs are large unless otherwise specified. Uncooked or partially cooked eggs should not be served to the very old, frail, young children, pregnant women or those with compromised immune systems.

First published in the UK in 2008
by Ryland Peters & Small
20–21 Jockey's Fields
London WC1R 4BW
www.rylandpeters.com

10 9 8 7 6 5 4 3 2 1

ISBN 978 1 84597 595 1

A catalogue record for this book is available from the British Library.

Printed and bound in China

contents

06 introduction
08 shake-and-serve dressings

10 fruit oils
12 extra virgin olive oil
26 argan oil
30 avocado oil

36 nut & seed oils
38 walnut oil
39 hazelnut oil
48 sesame oil
54 pumpkin oil
55 macadamia oil

62 vinegars
64 balsamic vinegar
70 vincotto
74 sherry vinegar
80 red and white wine vinegars
84 asian vinegar
90 apple cider vinegar
92 flavoured vinegars

95 suppliers
96 index

introduction

When I was a young girl, olive oil was something that came in tiny bottles from the chemist. It was considered a medicinal thing; something to do with unblocking ears. Very occasionally it was rubbed through our hair and left for a while – there was talk of it being a magical ingredient that would make hair ultra-shiny. Either way, no one would ever have dreamed of eating it.

It was a similar story with vinegar. Years ago, it came in two colours: brown and white. Brown was malt vinegar and we sprinkled it on our chips or stirred it into the bubbling pans of home-made toffee that we made once or twice a year as a treat. White was for pickling little onions, thin slices of freshly boiled beetroot, fine shreds of crunchy red cabbage or other such goodies from the garden or market. Sometimes it was used for cleaning; I remember my mum used to swear that adding a splash of it to water made the windows gleam.

A lot has changed since then. Today, medical research has confirmed that certain oils and vinegars have fantastic health-giving properties, and all around the globe, home cooks and professional chefs are beginning to discover just how versatile and exciting these ingredients are. In fact, they can transform a dish into something quite, quite special – in such an easy, uncomplicated and healthy way. A slick of oil over a piece of grilled fish, a light dressing over crisp salad leaves, a splash of vinegar in a marinade; such simple ideas with such stunning results. I've made cakes and cookies that would delight even the most ardent of dessert fans; and yet, using the fabulous monounsaturated oils in this book, cakes, cookies and pastries are no longer loaded with harmful saturated fats, and the results are amazing. Even though I have long used oils and vinegars in some shape or form almost every day, writing this book has given me the chance to experiment, indulge and enjoy them even more.

We have such lovely oils and vinegars available to us now, each with their own unique and delicious features. I hope this book will encourage you to make them your new culinary companions, and that you will love having them as part of your repertoire every bit as much as I do.

shake-and-serve dressings

Simply shake everything together in a screw-top jar and serve!

basic vinaigrette

8 tablespoons extra virgin olive oil

2 tablespoons white wine vinegar

1 teaspoon Dijon mustard

1 teaspoon caster sugar

sea salt and freshly ground black pepper

great with peppery leaves, spinach, slaws

walnut and blackberry vinegar dressing

6 tablespoons walnut oil

2 tablespoons blackberry vinegar (or to taste)

1 teaspoon caster sugar

red wine vinaigrette

2–3 tablespoons red wine vinegar

8 tablespoons extra virgin olive oil

freshly squeezed juice of ½ lemon

1 teaspoon caster sugar

2 shallots, peeled and chopped

1 garlic clove, crushed (optional)

sea salt and freshly ground black pepper

lemon, honey and thyme dressing

freshly squeezed juice of ½ small lemon (or to taste)

1 tablespoon runny honey

1 teaspoon fresh thyme leaves

8 tablespoons extra virgin olive oil

great with sweet leaves, noodles, fruit-based salads (Chinese leaves, pea shoot, melon)

mango, hazelnut and ginger dressing

2–3 tablespoons mango vinegar

8 tablespoons hazelnut oil

2 cm fresh ginger, grated

2 teaspoons runny honey

spiced argan, lime and honey dressing

6 tablespoons argan oil

freshly squeezed juice of ½ lime

2 teaspoons runny honey

2 teaspoons toasted cumin seeds, roughly ground

sea salt and freshly ground black pepper

toasted sesame, lemongrass and ginger dressing

2 tablespoons toasted sesame oil

1 lemongrass stalk, finely chopped

2 cm fresh ginger, grated

5 tablespoons extra virgin olive oil

2 tablespoons rice wine vinegar

1 teaspoon caster sugar

great with fish

fennel and chilli dressing

6 tablespoons Fennel and Fresh Lemon Oil (page 60)

a pinch of chilli flakes (or to taste)

freshly squeezed juice of ½ lemon

2 teaspoons caster sugar

hazelnut and orange dressing

8 tablespoons hazelnut oil

grated zest and freshly squeezed juice of 1 small unwaxed orange

1 tablespoon chopped fresh dill or chervil

sea salt and freshly ground black pepper

fruit oils

Long live olive oil! This magical oil is now an essential item in anyone's kitchen and has been lauded as a key player in the longevity of the Mediterranean peoples. And indeed it is a fantastic ingredient, as you will discover in the following pages. But there is so much more to oil than just extra virgin!

More and more supermarkets are stocking other fruit oils on their shelves and opening up the possibilities for home cooks around the country. Trendy argan oil is popping up following the recent popularity of Moroccan cuisine, and being used for more than just tagines; while the gorgeous emerald avocado oil shows just how precious the avocado is as a delicious, nutritious fruit.

A beautifully moist, moreish cake baked with avocado oil instead of butter? You'll be amazed at what you can do with fruit oils.

* extra virgin olive oil

Fossilized olive trees discovered during archaeological excavations of Bronze Age sites confirm the presence of wild olive trees as long ago as 12000 BC. But while historians continue to argue over the exact origins of the olive tree, cultivation is believed to have started around the vast, abundant plains of Mesopotamia around 5000 BC. From there, cultivation spread upwards and westwards through Egypt, Turkey, Greece, Italy and beyond. What goes wholly undisputed is that the olive and its oil have always been prized as a symbol of health, strength, peace and prosperity.

Thousands of years on, olive oil remains a hugely significant part of the Mediterranean diet, a diet which has become synonymous with good health. Concentrated scientific research undertaken over the past 30 years has confirmed that the olive is a powerhouse of nutrients, and that the oil extracted from the fruit shares all those remarkable qualities. The rest of the world seems to be catching on to the benefits of this wonderful fruit. Olive cultivation has now widened to include subtropical regions of Australia, New Zealand, South Africa, the USA, Chile and Argentina.

From the tree to the bottle

All olives begin life as a green fruit with a single stone. The flesh is made up of sugars and acids. As the fruit ripens on the tree, these acids and sugars are converted to oil, and the olive changes colour from green through to violet and then black. Hence black olives have a higher oil content and a very different flavour to that of green olives.

The olives are traditionally picked by hand. It's a labour-intensive and exhausting process, but any damage to the olives will result in an inferior oil. Nowadays machinery is available, but many producers still consider that the old-fashioned way produces a better quality of oil. Ideally, the olives should go from tree to press within 12 hours, to save unnecessary oxidation and damage to the fruit.

The olives are crushed mechanically. The oil is then separated from the paste by means of centrifugal force (i.e. by whizzing the paste around at high speed until the oil separates). Roughly 5 kg of olives are needed to produce just 1 litre of oil.

The blurb on the bottle

All virgin olive oils must meet strict criteria: they must be obtained solely by mechanical or physical measures, without the use of any chemicals, and processed under thermal conditions which do not lead to changes in the oil. The quality is then further classified according the 'organoleptic' properties (i.e. taste and aroma) and the presence of free acidity (this refers to the proportion of free fatty acids which are released when the olives are inferior or damaged in some way).

The four grades of edible olive oil are as follows:

Extra virgin olive oil
The highest quality rating, denoting oil which has perfect flavour and aroma and a free acidity level of less than 1 per cent.

Virgin olive oil
Virgin oils with barely detectable flavour and aroma defects and a free acidity of less than 2 per cent.

Semi-fine olive oil
Virgin oils with notable flavour and aroma defects and a free acidity level of less than 3.3 per cent.

Olive oil
Oils that are a blend of chemically refined and unrefined virgin oils. This grade represents a high proportion of the olive oil sold on the world consumer market. Blends are created to fit particular styles and prices; many of the oils labelled as 'extra light' are likely be a blend dominated by refined olive oil.

A question of health

When the Ancient Greek poet Homer labelled olive oil 'liquid gold', he wasn't too far off the mark. In nutritional terms, it's a superhero, rich in a whole gamut of health-giving vitamins, minerals and antioxidants. As with all the oils in this book, it is high in monounsaturated fats too. All very well, so long as you understand what that means! For better understanding, here's a quick explanation.

Saturated fats such as lard and butter are solid at room temperature and are mainly of animal origin. They are known to contain high levels of LDL (low-density lipoprotein) cholesterol – the 'bad' cholesterol that furs up arteries and causes cardiovascular disease. They should be avoided as much as possible.

Polyunsaturated fats are liquid at room temperature and most often found in foods of plant origin, such as sunflower and safflower oils. They are known to help lower levels of harmful LDL cholesterol.

Monounsaturated fats are also liquid at room temperature and found in foods of plant origin, such as olives, avocados and walnuts. Not only do these fats help to reduce levels of LDL cholesterol, but they are also known to increase levels of HDL (high-density lipoprotein) cholesterol, which is the 'good' cholesterol – so consider these monounsaturated fats as invaluable friends in the fight against dangerous clogged arteries.

Buying and storing oils

Heat and light both have a degenerative effect on oils. Always buy them in dark glass bottles, tins or ceramic containers.

Unlike many wines, oils do not grow old gracefully, so do not buy them in bulk. Ideally, once a bottle of extra virgin olive oil has been opened, it should be used within two or three months. Hopefully, after reading this book, you'll be struggling to keep a bottle anywhere near that long!

Frying with olive oil

Olive oil intermittently attracts bad press as a medium for cooking at high temperatures. Just like all fats, it will deteriorate during the frying process, especially if used repeatedly at high temperatures. Frying produces those dastardly do-badders, free radicals. However, because olive oil has a healthier composition than many of the usual frying oils, it is more resistant to harmful oxidization. So don't listen to those who try to put you off using olive oil for frying!

Looking at the label

First cold pressing/cold-pressed
This is a misleading phrase, often used for marketing purposes rather than as a guarantee of quality. During extraction, extra virgin olive oil must not be subjected

to heat above 30°C, but in any case, any olive oil producer worth his salt would never put the paste residue from a first pressing through his mills again in order to produce oil for culinary purposes, so the term 'first cold pressing' or 'cold-pressed' is unnecessary.

Single-estate and estate-bottled oils
These are first-class oils from olives grown on a single estate, usually picked by hand, pressed and bottled on site. They are among the best olive oils available and this will usually be reflected in the price.

Filtered and non-filtered
Traditionally, olive oils were often cloudy, with residue from the pressing remaining in the oil. Improved filtering techniques have resulted in clearer oils being considered the norm. Today, non-filtered oils are frequently the wares of devoted artisan producers who believe that less interference creates better oil, and the quality is therefore often excellent.

DO, DOP, DOC, PGI and PPO
These abbreviations are marks of quality and relate to an oil's 'designation of origin'. The mark is awarded to oil when every stage of the production process (including the olive harvest) has taken place within an elected geographical area. This is to guarantee precise standards and ensure a product maintains the typical characteristics, and adheres to the customs, of the district in which it is made.

Date of pressing
Sadly, this information is rarely found on bottles now. It seems to have been replaced in many cases by a 'best before' date, which isn't a clear indication of when the oil was pressed. Quality producers will be proud of their 'new season's' olive oil.

Around the world

When it comes to the individual flavour and characteristics of olive oils, of course there are many contributing factors. Olive variety, climate, weather, farming practices and production methods will all have an effect on the final result. That said, whenever I'm asked how I choose which oil to use for a particular dish, I tend to advise people to look first to the oil produced by the country the recipe comes from. For instance, a gutsy Greek roast pepper and feta dip will work well with a plucky, herby Greek oil; a full-flavoured Andalusian pork hash will dance with a punchy oil from olives grown under the region's intense sun; and the peppery oils of Tuscany complement such wonderful traditional rustic dishes as *pappa al pomodoro*.

As a very general rule of thumb, those regions in which the olives are exposed to harsh sun and dramatic variations in temperature will be inclined to produce more pungent oils. Areas with a more moderate climate will lean towards lighter flavoured oils.

Here are some characteristics and varieties of the olives particular to each oil-producing country or region.

Southern France/Provence
These oils usually display sweet, rounded flavours from flowery to mildly spicy. Good olive varieties: Picholine, Aglandou.

Italy
Italian oils are as diverse as the people themselves. Tuscan oils tend to have lovely peppery overtones; Umbrian oils can range from fruity and sweet to lightly piquant. Abruzzese oils share the peppery qualities of the Tuscan oils, but have a smooth fruitiness too. Further south, in Calabria, the olives are often left to fall naturally into waiting nets rather than being picked by hand, which produces a unique oil. Good olive varieties: Leccino, Moraiolo, Coratina, Frantoio.

Spain
Light, sweet oils from the north give way to feisty, assertive oils from the hot hills of Andalusia. Good olive varieties: Picual, Hojiblanca, Lechin, Arbequina.

Greece
Oils can be punchy, herby and sometimes slightly bitter. Good olive variety: Koroneiki.

New World
Relatively new kids on the block, Australia and New Zealand are producing some great oils with a range of styles and flavours – from buttery and sweet to woody and warm. Good olive varieties: a whole melting pot – Kalamata, Frantoio, Correggiola, Nevadillo Blanco.

Argentina
Argentina is currently the principal country producing olive oil in South America, largely due to national government incentives in certain regions. Similar climate and olive varieties to Spain tend to produce oils with similar characteristics. Good olive varieties: Criolla, Arbequina.

California
Another relatively recent addition to the world of oil producers, California now produces around 97 per cent of the USA's olive oil. As with Australia, they have taken some of the best olive varieties, producing oils with a wide range of characteristics, from light to full-bodied. Good olive varieties: Mission, Century Mission, Frantoio, Arbequina, Picholine.

hot garlic prawns with spinach and ladolemono sauce

This traditional Greek olive oil and lemon combination makes a lip-smacking sauce that works especially well with moist grilled fish and dark, peppery vegetables.

1 tablespoon extra virgin olive oil

2 garlic cloves, crushed

500 g uncooked king prawns, shelled and deveined

1 kg spinach

Sauce

100 ml extra virgin olive oil (Greek would be perfect!)

grated zest and freshly squeezed juice of 1 unwaxed lemon

a pinch of caster sugar

sea salt and freshly ground black pepper

Serves 4

To make the ladolemono sauce, whisk the olive oil, lemon zest and juice and caster sugar together until emulsified, then season to taste.

Heat the olive oil in a large frying pan or wok and add the garlic. Cook for 2 minutes over medium heat. Add the prawns and stir-fry for 3–4 minutes, until pink and cooked through.

Meanwhile, blanch the spinach in salted boiling water for 1–2 minutes, until just wilted. Drain and squeeze dry. Dress with the ladolemono sauce and serve immediately with the hot prawns. Pour over the garlicky juices from the frying pan.

black grape schiacciata

This is a gorgeous bread to serve warm from the oven at breakfast time. I've used fresh black grapes here, but at the cookery school I run with my dear friend Laura in Italy, we make a lovely version using semi-dried Montepulciano grapes from Laura's wine estate and serve it with salty cheeses such as Gorgonzola or Parmesan. If you wanted to try something similar, you could substitute semi-dried cherries or plump Lexia raisins.

100 ml fruity extra virgin olive oil

a large handful of fresh rosemary

450 g extra strong bread flour

1 teaspoon salt

2 tablespoons caster sugar

1 teaspoon quick-rise dried yeast

250 ml hand hot water

400 g seedless black grapes

a deep baking tin, 23 x 30 cm, oiled

Serves 6

Put the olive oil and rosemary in a bowl. Give the rosemary several good squeezes to release the aroma into the oil. Set aside for a few minutes.

Put the flour, salt and 1 tablespoon of the caster sugar in a large bowl and stir well. Add the yeast and stir again. Pour in 2 tablespoons of the infused olive oil and enough hand-hot water to bring the mixture together to a soft but not sticky dough (the exact amount of water needed will vary according to the flour used).

Turn the dough out on to a lightly floured surface and knead for 5 minutes, or until the dough is smooth and elastic. Fold in the grapes and knead for a further 2–3 minutes. The dough may become sticky at this point, so dust with a little extra flour if necessary.

Press the dough into the prepared baking tin and push it with your knuckles to fill the tin. Leave the dough to rise in a warm place for about 40 minutes, or until it has doubled in size. Preheat the oven to 220°C (425°F) Gas 7.

Drizzle the remaining olive oil over the risen dough, and scatter some of the rosemary leaves and the remaining caster sugar evenly across the top. Bake in the preheated oven for about 25 minutes, or until the surface is golden brown and the base sounds hollow when tapped. Leave to cool on a wire rack.

rosemary and roast potato tart with olive oil pastry

Using extra virgin olive oil in pastry works beautifully – the result is light, crisp pastry with a lovely flavour. Ultra-thin slices of potato pepped up with plenty of fresh rosemary make a simple but surprisingly delicious topping. The beauty is in the simplicity, but you could ring the changes by adding a crumbling of piquant blue cheese, goats' cheese or feta over the base before layering the potato on top.

400 g floury potatoes, peeled and thinly sliced

4 tablespoons extra virgin olive oil

3–4 sprigs of fresh rosemary

sea salt and freshly ground black pepper

Pastry

160 g plain flour

a pinch of salt

80 ml fruity (or slightly peppery if preferred) extra virgin olive oil

Serves 4

Preheat the oven to 180°C (350°F) Gas 4.

To make the pastry, sift the flour into a large bowl and add the salt. Add the olive oil and enough water to bring the mixture together to form a soft but not sticky dough. Wrap the dough in clingfilm and leave to rest for 30 minutes or so.

Put the potatoes in a large bowl and cover with cold water. Leave for 2–3 minutes to draw out the starch.

Put the olive oil and rosemary in a bowl. Give the rosemary several good squeezes to release the aroma into the oil. Set aside for a few minutes.

Drain the potatoes and dry well on kitchen paper. Toss them in the infused oil and season.

Roll out the pastry thinly on a baking sheet to form a circle 23 cm in diameter. Layer the potatoes evenly over the pastry.

Bake in the preheated oven for about 25 minutes, or until the potatoes are soft and golden. Serve hot with a crisp green salad.

QUALITE EXTRA

olive oil and orange cake

I was absolutely bowled over when I first discovered the fantastic results to be had from using oils rather than hard fats in cakes. And this cake is without doubt one of my absolute favourites – it's so moist and yet has a light-as-air texture, with a gorgeous orangey flavour. Perfect with a cuppa at any time, it makes a lovely dessert when served with summer berries too.

1 large unwaxed orange

1 large unwaxed lemon

100 ml fruity extra virgin olive oil

175 g caster sugar

4 eggs

175 g ground almonds

2 teaspoons baking powder

icing sugar, to dust

a 20-cm loose-bottomed cake tin, lightly oiled and base-lined with non-stick baking parchment

Serves 8–10

Wash the orange and lemon and put them both in a saucepan. Cover with water, bring to the boil and simmer for 30 minutes or so, until soft. Remove from the water and leave to cool. Cut the orange in half, discard the pips and put the skin and pulp in a food processor. Cut the lemon in half and discard the pips and pulp. Put the skin in the food processor with the orange, whizz to a purée and set aside.

Preheat the oven to 180°C (350°F) Gas 4.

Beat the olive oil, sugar and eggs together until light and fluffy. Stir in the ground almonds and baking powder. Add the puréed fruit and stir until thoroughly mixed. Spoon the batter into the prepared cake tin.

Bake in the preheated oven for 50–60 minutes, or until the cake is golden and risen and springs back when touched with a fingertip. Leave to cool in the tin until completely cold. Turn out and dust with icing sugar. Serve in slices with fresh summer berries, if liked.

*argan oil

A first visit to the parched and scrubby Argan Forests of south-west Morocco may have you seriously questioning your sanity – or at least your eyesight. The gnarled, thorny trunks of the argan tree allow a nimble-footed breed of goat to clamber along its branches, where they can be spotted precariously balanced at the top of the tree, munching happily on argan leaves and fruit.

The argan fruit has a green and fleshy appearance, not unlike a large unripe olive. Tucked inside is a tough-shelled nut containing up to three kernels. When the goats eat the fruit, the outer fleshy part is easily digested but the nuts aren't. As you might expect, nature takes its course, after which the undigested nuts are collected from the ground by Berbers and pressed to produce delicious and distinctive argan oil!

Sadly, argan oil is still relatively expensive. Production continues to be on a fairly small scale, much of it still using conventional methods. Traditionally, Berber women cracked open the nuts by hand – a slow process. It took up to 20 hours simply to render enough kernels to make just a single litre of oil. The oil was then extracted using a press powered by animals; and not much has changed today.

Argan oil has a faint reddish tinge and a pleasant nutty aroma. In Morocco, it is typically used as cooking oil, or to stir into couscous or add a final flourish to tagines. The residual substance after extraction is a thick chocolate-coloured paste called *amlou*, which is sweetened with honey and served with bread for breakfast. I sometimes mix ground almonds, honey and argan oil together, which produces a similar spread and is delightful on hot toast.

As with all of the other oils mentioned in this book, argan oil is high in monounsaturated fats, essential fatty acids and antioxidants. It is now becoming very trendy in Middle Eastern restaurants all over the world.

As well as the traditional Moroccan uses for the oil I have mentioned, it does make a lovely dressing for salads and vegetables. I love its unique flavour and serve it unadorned poured over crisp salad leaves, maybe with a touch of salt and freshly ground black pepper. Although it has a reputation for being the most costly oil in the world, a little does go a long way – and despite its rather unappealing beginnings, it really is a wonderful product!

lamb and butternut squash tagine

When you are making this tagine, reserve the butternut seeds, rinse them, toss them with a little argan oil and roast them, then stir them into the couscous with a handful of raisins. Do as the Moroccans do and stir a little oil into the couscous too – it's fabulous.

800 g stewing lamb, cubed

3 tablespoons plain flour seasoned with salt and pepper

1 teaspoon cumin seeds

1 teaspoon fennel seeds

6 tablespoons argan oil

1 onion, chopped

1 small butternut squash, peeled, deseeded and cubed

600 ml lamb or vegetable stock

a pinch of saffron threads

1 cinnamon stick

1 whole garlic bulb, separated

a bunch of fresh coriander, chopped

Serves 4

Place the lamb in a plastic bag with the seasoned flour and shake to coat well.

Toast the cumin and fennel seeds for 1–2 minutes in a small, dry frying pan over medium heat, until wonderfully fragrant.

Heat 2 tablespoons of the argan oil in a large casserole, then fry the onion for 2–3 minutes, until soft. Add the lamb and cook for a further 4–5 minutes, until golden all over. Add the butternut squash and cook for 3–4 minutes.

Pour in the stock and add the toasted spices, saffron and cinnamon. Add the separated garlic cloves (there is no need to peel them). Bring to the boil, then turn down the heat and simmer for about 1 hour 30 minutes, or until the lamb is tender.

Stir in the coriander. Serve immediately, with couscous or rice and the remaining argan oil drizzled over it. Do encourage diners to squeeze the softened garlic out of its skin and smear it on the tender chunks of lamb. Mmmmmm.

* avocado oil

The avocado pear is native to Mexico and Central America. Archaeologists have discovered fossilized avocado seeds in Mexico which may date as far back as 6000 BC. The Aztec name for the avocado was *ahuacatl*, or 'testicle', a name given to the fruit because it hangs in pairs on the tree and was thought to resemble that part of the male anatomy. Young Aztec maidens were even forbidden to go out at harvest time. But as well as being celebrated for its aphrodisiac properties, the avocado was also highly favoured as a foodstuff. Guacamole, the ever popular avocado dip, is thought to be based on the Aztec speciality *ahuaca-mulli*, a sauce prepared from mashed avocados.

In 1519, when a Spanish soldier of fortune named Hernando Cortez conquered Mexico, he discovered that the avocado formed a fundamental part of the native diet. But the clever and resourceful conquistadors also discovered that the milky fluid secreted by the central seed turned red when oxidized, making an indelible ink suitable for writing. Documents written in avocado ink are still available today.

The oil of avocados has long been known for its exceptional properties as a beauty aid in cosmetics and hair preparations; but now it is generating huge excitement in the kitchens of top chefs around the world. And rightly so: it's amazing stuff! Made from the flesh of the avocado rather than the seed, it has a stunning emerald green colour, and the unmistakable buttery, rich flavour of avocado.

The fruits are left to grow on the tree for around 18 months, after which they are picked, graded and transferred to ripening houses where they are left to mature. The skin and seeds are then removed. The oil is extracted from the pulp, filtered and bottled.

Cholesterol-free, rich in monounsaturated fats and oozing with antioxidants, vitamins and minerals, avocados are currently considered one of the healthiest fruits on earth. It goes without saying that the cold-pressed oil has the same health properties.

Avocado oil also has an unbelievably high smoke point, making it suitable for cooking at high temperatures and ideal for stir-frying, but I think it comes into its own for dipping, drizzling and dressings. As you might imagine, it's especially good stirred into dishes that feature avocados – it amplifies the flavour of dips such as guacamole and is lovely stirred into soups and tossed into salads. Add a slick of oil to a salad of juicy tomatoes, sliced avocado and roughly torn buffalo mozzarella and finish off with a good grinding of black pepper and a few leaves of young basil; bliss. Or try drizzling the oil over steamed tenderstem or purple sprouting broccoli. Use it as an alternative to olive oil in salsa verde – it makes a fabulous accompaniment to roast chicken or grilled fish.

It's also a lovely oil to serve alone with raw vegetables and chunks of bread for dipping in it but my most exciting discovery came when I used it to replace butter in cakes – try it and see for yourself.

Italian vegetable and bread soup

This meal-in-a-bowl soup is loosely based on the Tuscan vegetable and bread soup ribollita, but this version illustrates how beautifully avocado oil complements pulses and vegetables. It marries particularly well with dark green vegetables. It's one of those recipes that has three essential elements – good oil, fresh vegetables and good bread. The rest is interchangeable, according to the seasons and your own fancy.

6 slices of ciabatta bread

about 80 ml avocado oil

2 red onions, chopped

2 sticks of celery, chopped

2 carrots, chopped

2 garlic cloves, crushed

6 ripe tomatoes, deseeded

1½ litres well-flavoured vegetable stock

two 400-g tins cannellini beans, drained and rinsed

4 courgettes, sliced

200 g kale, chopped

100 g savoy cabbage, shredded

freshly ground black pepper

Serves 4

Preheat the oven to 200°C (400°F) Gas 6.

Drizzle the ciabatta slices with a little of the avocado oil and bake in the preheated oven for 5–6 minutes, or until crisp. Remove from the oven and set aside.

Meanwhile, heat 3 tablespoons of the avocado oil in a large saucepan and fry the onions, celery, carrots and garlic over low heat for 4–5 minutes, or until the vegetables are shiny and starting to soften. Add the tomatoes and cook for a further couple of minutes.

Pour in the stock and cook for 15 minutes. Add the beans and cook for a further 5 minutes. Add the courgettes, kale and savoy cabbage and cook for 4–5 minutes more, or until the greens are just cooked but still retain their colour.

Break the ciabatta into bite-sized pieces and divide equally between four warmed soup bowls. Spoon the soup into the bowls and add a good grinding of pepper. Drizzle with a little more avocado oil and serve immediately.

avocado oil, lemon and pistachio cake

This is without doubt a very special cake – one of my absolute favourites and one which has surprised and delighted those who have tasted and tested!

150 ml avocado oil

200 g caster sugar

3 eggs, beaten

150 g shelled pistachios

100 g ground almonds

grated zest and freshly squeezed juice of 2 unwaxed lemons

50 g plain flour

2 teaspoons baking powder

100 g granulated sugar

a 20-cm loose-bottomed cake tin, lightly oiled and base-lined with non-stick baking parchment

Serves 8–10

Preheat the oven to 170°C (325°F) Gas 3.

Beat the avocado oil, caster sugar and eggs together until light and fluffy.

Chop 100 g of the pistachios very, very finely and add to the egg mixture along with the ground almonds, lemon zest, flour and baking powder. Spoon the batter into the prepared cake tin and level the surface. Coarsely chop the remaining pistachios and scatter evenly over the surface of the cake.

Bake in the preheated oven for 40–45 minutes or so, or until the cake has risen and springs back when touched with a fingertip.

Mix the lemon juice and granulated sugar together to make a syrup. Remove the cake from the oven and pour the syrup evenly over the cake while it is still hot.

nut & seed oils

When buying nut and seed oils, avoid oils that have been extracted using chemicals (especially hexane, during which process the high temperatures can damage oil quality). Choose mechanically pressed oils instead and store them in a cool place. Refrigerate once opened.

All oils react differently to being heated, so they are measured to find the maximum temperature they can safely be heated to. This is the 'smoke point' (i.e. the point at which smoke is first detected when it is heated in a specially lit, draught-free laboratory) and it determines whether an oil is best used to dress a dish, to shallow-fry at a low heat, or for frying at high temperatures.

* walnut oil

Archaeologists in south-west France have uncovered petrified walnut shells dating back 8,000 years. Inscriptions on clay tablets found in Mesopotamia record the existence of walnut groves in the Hanging Gardens of Babylon. The Ancient Greeks believed the walnut looked so much like a human brain that it must have magical and mystical properties. In fact they extracted the oil from walnuts even before they began to make olive oil.

The Périgord region in France is renowned for producing some of the best-quality walnuts in the world, so the oil from this region has long been deemed the finest. However, following the first commercial planting of walnut trees in California in 1867, California now contributes two-thirds of the world's supply of walnuts and the quantity and quality of oil production is continually improving.

Premium-quality walnut oil has a lovely topaz colour, a wonderful rich flavour and all the health benefits of being cholesterol-free and high in monounsaturated and polyunsaturated fats. It is also high in omega-3 and omega-6 essential fatty acids, vitamins and minerals.

During production the harvested nuts are graded and dried, then the kernels are ground and roasted in cast-iron kettles. The warm paste is transferred to a hydraulic press, where the oil is extracted mechanically before being filtered and bottled. Some cheaper oils are produced by simply macerating the nuts in vegetable oil, which gives an inferior flavour and a lower nutritional content.

Although walnut oil has quite a high smoke point, its delicate flavour is lost with excessive heat, so the oil is far better suited to being used as a dressing for salads, or served over freshly cooked fish, pasta or meat. It has a particular affinity with fish, bitter leaves such as chicory, and salads containing citrus fruits.

For a delicious and refreshing salad, try a splash of walnut oil and a sprinkling of sherry vinegar over thinly sliced oranges, red onions and rocket.

Root vegetable soups taste wonderful served with a swirl of walnut oil. Drizzle the oil over cubes of bread and bake them to create crunchy, walnut-flavoured croutons for a fabulous accompaniment. Walnut oil also makes a great match for vegetables belonging to the brassica family (e.g. cabbage, Brussels sprouts, cauliflower). The smallest trickle and simply steamed vegetables take on a new life. Play around with cakes, cookies and pastries too – the results are fantastic.

*hazelnut oil

Hazelnuts were first discovered growing wild along the Black Sea coast as early as 300 BC. The commercial cultivation of hazelnuts has been the primary source of income for the region for centuries and continues to be so today. Turkey provides 70–75 per cent of the total world crop, with Europe, China, Australia, California and other parts of the USA making up the rest.

Throughout history, the hazelnut and its branches have been associated with enchantment and mystery. Any self-respecting fairy godmother would have had a hazel branch wand. By tradition, hazel branches have always been used as rods for water divination. Prior to the seventeenth century, not only was it believed that they could help to find water, but they could also allegedly assist in tracking down murderers and thieves – and even, occasionally, the odd chest of missing treasure. A hazelnut tucked inside a pocket was believed to bring wealth, security and good luck, while a whole string of hazelnuts hung inside the house was thought to do the same thing on a larger scale. Eating them apparently bestowed knowledge and boosted fertility.

Although an astonishing 80 per cent of the world's total annual hazelnut crop is used in the chocolate industry, happily some of the remainder goes into the production of hazelnut oil. A beautiful, elegantly flavoured oil (very popular in French cuisine), it is high in monounsaturated fat, cholesterol-free and indeed shares many of the health benefits of extra virgin olive oil.

The oil is produced in much the same way as walnut oil. The nuts are dried after harvesting, then ground and roasted. The paste is then pressed in a hydraulic press and the oil is filtered and bottled. Its fine flavour is easily diminished by cooking at high temperatures, and so it shines in salads and as a dressing for lightly cooked spring vegetables, pulses and pasta. I have a particularly soft spot for hazelnut oil dribbled over jacket potatoes or trickled over roasted asparagus, as a substitute for butter.

For those with a sweet tooth, hazelnut oil gives a lovely light touch to cakes and cookies, imparting a gentle but unmistakable hint of hazelnuts. It makes heavenly feather-light hazelnut friands to accompany fresh summer berries and white chocolate mousse, or deliciously different crisp wafer cones to fill with cappuccino ice cream.

roast onion and celeriac ravioli with warm walnut pesto

Walnuts and celeriac make a magical combination, especially when the celeriac is roasted with onion and tucked inside ravioli, as it is here. However, if time is short, try this lovely pesto tossed through one of the tubbier pasta shapes such as rigatoni, radiatore or penne.

sea salt and freshly ground black pepper

semolina, to dust

Pasta

400 g Type 00 pasta flour

4 eggs, beaten

Filling

1 celeriac, peeled and cubed

1 large onion, finely chopped

1 garlic clove, peeled and crushed

1 teaspoon fresh thyme leaves

2 tablespoons honey

3 tablespoons extra virgin olive oil

Pesto

100 g walnuts

2 garlic cloves, peeled and crushed

3 tablespoons chopped fresh rosemary

50 g pecorino Romano cheese, grated

4 tablespoons walnut oil

Serves 4

To make the pasta, sift the flour into a bowl or food processor. Add the eggs and bring the mixture together to make a soft but not sticky dough. Turn out on to a lightly floured surface and knead for 4–5 minutes, until smooth. Wrap in clingfilm and refrigerate for at least 30 minutes. Preheat the oven to 200°C (400°F) Gas 6.

To make the filling, put the celeriac, onion, garlic and thyme in a roasting tin. Drizzle with the honey and olive oil and some seasoning. Toss well. Roast in the preheated oven for 25 minutes, or until soft and golden. Leave to cool slightly, then mash coarsely.

To make the pesto, chop the walnuts, garlic and rosemary very finely in a food processor or by hand. Stir in the pecorino and walnut oil. Season to taste.

Roll out the chilled pasta to a thickness of 1 mm using a pasta machine (or a rolling pin and plenty of elbow grease). Cut the pasta into two long equal pieces. Place teaspoonfuls of the filling at even intervals across one half of the pasta. Brush around the filling with a little water and cover with the second sheet. Press lightly around the filling to seal, then cut into squares using a sharp knife or pastry wheel. Lay out the ravioli on a sheet of greaseproof paper lightly dusted with semolina.

Bring a large saucepan of salted water to the boil and drop in the ravioli. Cook for 2–3 minutes, until the pasta rises to the surface and is soft but still retains a little bite. Drain and toss immediately with the pesto. Serve at once.

griddled lamb fillet with garlic and walnut aïoli

This walnut-scented, garlicky mayonnaise is dangerously addictive. Not only is it great with lamb, but it's also fantastic for dunking crisp roasted potato wedges into!

3 tablespoons extra virgin olive oil

2 tablespoons thyme leaves

800 g lamb fillet

sea salt and freshly ground black pepper

Aïoli

1 egg

2 tablespoons white wine vinegar

1 tablespoon Dijon mustard

200 ml sunflower oil

100 ml walnut oil

3 garlic cloves, peeled and crushed

Serves 4

To make the aïoli, put the egg, vinegar and mustard in the bowl of a food processor (or use a hand-held blender). Add a good pinch of salt and a touch of black pepper. Set the motor running, and slowly add the sunflower oil. The mixture should start to emulsify. When all the sunflower oil has been incorporated, slowly add the walnut oil. The mixture should be glossy and thick. Add the garlic and check the seasoning (you may want to add more black pepper here).

Mix the olive oil and thyme together in a shallow bowl. Cut the lamb into slices and toss in the oil to coat, then season.

Heat a ridged griddle pan until very hot and cook the lamb for 2–3 minutes on each side, or a little longer if you prefer the lamb to be well done.

Serve the lamb and aïoli together. Asparagus or roast vegetables make great accompaniments.

marinated chicken, raisin and chilli salad with hazelnut dressing

Everyone seems to love this moist marinated chicken salad. For the best flavour, be sure to toast the hazelnuts until dark golden. If time is short, you could cheat a little and buy a freshly cooked spit-roasted chicken and throw away the evidence before anyone sees...

a 2-kg free-range chicken

1.5 litres good chicken or vegetable stock

a handful of raisins

100 g blanched hazelnuts, toasted until dark golden

1 teaspoon dried chilli flakes, or to taste

a small bunch of fresh flat leaf parsley, chopped

sea salt and freshly ground black pepper

Hazelnut dressing

6 tablespoons hazelnut oil

2 tablespoons red wine vinegar

1 tablespoon caster sugar

Serves 4–6

Put the chicken in a large saucepan and cover with the stock. Bring to the boil. Turn down the heat and poach the chicken for about 1 hour, or until the juices run clear when the chicken is tested with a skewer at the thickest part of the leg. Leave to cool in the stock, then transfer to a chopping board. Using a rolling pin, give the chicken a few good thwacks along the breast. This will make it much easier to shred. Shred the meat into a large bowl. Add the raisins, hazelnuts and chilli flakes.

To make the hazelnut dressing, stir the hazelnut oil, vinegar and sugar together and season to taste. Add to the chicken along with the parsley, toss well and leave to marinate in a cool place (not the fridge) for at least 1 hour. Serve at room temperature.

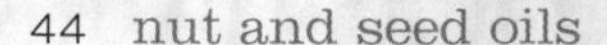

fig and hazelnut breakfast bread

I love this bread toasted for breakfast. Rather than spreading it with butter, I often serve it with a sort of dressing which I make by whisking a little hazelnut oil and honey together with some orange zest, or the simple Sweet Lemon and Olive Oil Conserve below. The bread is also lovely served with cheese.

500 g strong bread flour

50 g caster sugar

2 teaspoons salt

a 7-g sachet fast-action dried yeast

2 tablespoons hazelnut oil

200 ml hand-hot water

2 tablespoons blanched hazelnuts, toasted and chopped

300 g ready-to-eat dried figs, quartered

Makes 2 loaves

Sift the flour, sugar and salt into a bowl and stir well. Add the yeast and stir again. Pour in the hazelnut oil and enough hand-hot water to bring the mixture together to a soft but not sticky dough (the exact amount of water needed will vary according to the flour used). Add the hazelnuts and figs and knead well to combine.

Turn the dough out on to a lightly floured work surface and knead for 5–10 minutes. Divide into two rough loaves and put on oiled baking trays spaced well apart. Leave to rise in a warm place for 1 hour, or until the dough has doubled in size. Preheat the oven to 220°C (425°F) Gas 7.

Cut a couple of slashes in the top of each loaf and bake in the preheated oven for 25–30 minutes, until golden and firm and the bases sound hollow when tapped. Leave to cool on a wire rack.

sweet lemon and olive oil conserve

50 ml extra virgin olive oil

150 g caster sugar

grated zest and freshly squeezed juice of 2 unwaxed lemons

4 egg yolks

Put the olive oil, caster sugar, lemon zest and juice and the egg yolks in a microwave-proof bowl and mix to combine. Microwave on full power for 3–5 minutes, stirring every minute, until the curd has thickened. Leave to cool (as it cools it will thicken more). Spoon into clean, sterilized jars and refrigerate until ready to use.

sesame oil

Records show that sesame seeds have been around since prehistoric times, but according to Assyrian myth, the seeds go back even further – it is believed that the gods drank sesame wine the night before they created the earth.

You might think that scattering sesame seeds on bread before baking is a fairly recent innovation, but history tells us otherwise. The decorated ancient tomb of one Egyptian noble, dating back 4,000 years, depicts a baker adding sesame seeds to his bread dough.

Sesame oil has been around a long time too: at least 5,000 years. From as early as 2100 BC, the Babylonians were using sesame oil to cook with, to make wines and brandy, and as a base for perfumes. The Chinese used the oil as lamp fuel, and created the soot from which their exclusive stick ink was made. In museums today you will find ancient Chinese scripts using ink made from sesame oil that date back around 5,000 years.

Sesame oil is cholesterol-free, high in monounsaturated and polyunsaturated fats and contains notable amounts of magnesium, copper, calcium, iron and vitamin B_6.

Dark sesame oil is generally the oil to use in a recipe from China or Japan. It has a deep amber colour and pronounced sesame flavour. This is because the seeds are hulled and toasted before pressing. However, it does have a low smoke point and will burn easily if used at high temperatures (unless blended with other oils more suited to frying), so is usually used sparingly to add flavour after cooking.

Oil pressed from raw, untoasted seeds produces very light, almost colourless sesame oil with little flavour. It is commonly found in Indian cuisine, where it is known as gingelly or til oil.

A dash or two of sesame oil adds a typical taste of the orient to stir-fry dishes – one of the favourite snack foods of my youngest son, Tim, is a simple bowl of freshly cooked noodles with a drizzle of toasted sesame oil. It also makes a great addition to home-made pitta bread and flour tortillas.

Sesame oil, like some of the oils already mentioned in this chapter, adds a special touch to baked items, but especially to breads and biscuits. Personally, I love it teamed with dark berry and stone fruit vinegars such as blackberry and plum, to make punchy dressings for crispy duck salad and goose or pork rillettes. For a delicious variation on the traditional Middle Eastern dip hoummus, substitute natural yoghurt and a spoonful of toasted sesame oil for the customary ground sesame paste, tahini.

a bit like bang-bang chicken

This is just as the title says – a bit like bang-bang chicken! To save time, you could cheat and use a bought, ready-roasted chicken or even smoked chicken breasts. I love to substitute ready-cooked duck from a favourite Chinese restaurant, too.

Salad

a 2.5-kg free-range chicken

1.5 litres chicken or vegetable stock

3 carrots, peeled and shredded

1 small cabbage, peeled and shredded

3 mixed peppers, cut into strips

100 g sugar snap peas, diagonally halved

3 tablespoons toasted sesame oil

50 g sesame seeds

Sauce

200 g peanut butter

3–4 tablespoons Easy Chilli Jam (see page 81) or shop-bought chilli sauce

2 cm fresh ginger, peeled and grated

3 tablespoons toasted sesame oil

2 tablespoons extra virgin olive oil

Serves 4

Put the chicken in a large saucepan and cover with the stock. Bring to the boil. Turn down the heat and poach the chicken for about 1 hour, or until the juices run clear when the chicken is tested with a skewer at the thickest part of the leg. Leave to cool in the stock, then transfer to a chopping board. Using a rolling pin, give the chicken a few good thwacks along the breast. This will make it much easier to shred. Shred the meat into a large bowl.

Mix together the carrots, cabbage, peppers and sugar snap peas in four individual bowls. Top with the shredded chicken.

Heat the sesame oil in a small frying pan and fry the sesame seeds until golden.

To make the sauce, put the peanut butter, chilli jam, ginger, sesame oil and olive oil in a saucepan, whisk to combine and heat gently over low heat.

Drizzle some of the sauce over the chicken and scatter with the sesame seeds. Serve immediately, with the remaining sauce alongside.

sesame and rice vinegar wafers with mango and coconut dips

I love these wafers – they make the lightest, prettiest and tastiest accompaniment to mousses, ice creams and dips such as the lovely two-coloured one here.

Wafers

2 tablespoons jaggery (or demerara sugar)

1 tablespoon rice vinegar

1 tablespoon toasted sesame oil

2 tablespoons sesame seeds

2 tablespoons desiccated coconut

1 tablespoon plain flour

a pinch of salt

Dips

1 ripe mango, peeled, stoned and chopped

1 tablespoon rice vinegar

3 tablespoons caster sugar (or to taste)

200 ml coconut cream

120 ml Greek yoghurt

a baking sheet, lined with non-stick baking parchment

Serves 4

Preheat the oven to 170°C (325°F) Gas 3.

To make the wafers, put the jaggery and rice vinegar in a small saucepan and cook over low heat (or microwave on full power for 30–40 seconds) until the jaggery has melted. Stir in the sesame oil, sesame seeds, coconut, flour and a pinch of salt. Leave to cool and firm up slightly.

Place teaspoonfuls of the mixture on the prepared baking sheet, spacing them well apart to allow the mixture to spread. Bake in the preheated oven for 4–5 minutes, until golden. Remove from the oven and carefully drape the wafers over a lightly oiled rolling pin. Leave until completely cold and store in an airtight container until ready to use.

To make the dips, put the mango, rice vinegar and 1 tablespoon of the sugar in a food processor and whizz to a purée. Transfer to a small jug.

Put the coconut cream, yoghurt and remaining sugar in a separate small jug and stir to combine. To serve, slowly pour the dips simultaneously into a large bowl (or use four smaller individual bowls), taking care to pour the dips on opposite sides of the bowl. Swirl the centre with a fork to create a pattern. Serve with the wafers.

*pumpkin oil

Pumpkins are believed to have originated in Central America. Seeds from associated plants have been found in Mexico from as long ago as 5500 BC. Pumpkins were prized for their flesh and seeds, and the shells were used as serving pots.

Long before the arrival of the Pilgrim Fathers in America, pumpkins formed an integral part of the diet of Native American Indians. When the white settlers arrived, pumpkins became a key ingredient in their diets, too. Pumpkins were popular in soups, stews and sweet dishes. Around this time, the first pumpkin pie was born – the seeds were removed and the fleshy shell was filled with milk and honey and flavoured with spices. When baked, the flesh became soft and was then scooped out into bowls to be eaten. The shells were dried and cut into strips to weave into carpets.

From America, the pumpkin found its way across to Europe, where it gained popularity. Despite this, it is only grown as an oil plant in a certain region of central Europe: a small intersection comprising parts of Austria, Slovenia and Hungary. Since Austria joined the European Union, the market for pumpkin oil has increased and in turn so has production, but most of the oil currently comes from Styria, in south-east Austria.

Pumpkin oil has a dark, concentrated green colour, gorgeous velvety texture and intense nutty flavour. It is high in monounsaturated fats, provides an ideal blend of omega-3 and omega-6 oils and is rich in many vital nutrients, including vitamins A and E, selenium and zinc.

The seeds are coarsely ground, then toasted at temperatures of around 60°C before the oil is extracted.

The oil has a fairly low smoke point, so is best reserved as a salad oil and to sprinkle over cooked dishes. It makes a delicious alternative to butter on baked or roasted root vegetables, and is excellent drizzled over soups and casseroles. Combined with warm, sweet spices such as cinnamon and nutmeg, it's a marriage made in heaven. It has a natural affinity with chocolate, too – it makes a particularly fabulous, glossy chocolate rum sauce and the most amazing chocolate nutmeg brownies. Whenever I make them, the sauce gets slurped up and the brownies disappear in no time – and when I reveal that the secret ingredient is pumpkin oil, everyone is always amazed!

* macadamia oil

Macadamia nuts have grown in Australia's rainforests for thousands of years. They formed an important part of the diet eaten by Aboriginal tribes, whose typical victuals were quite low in fats and oils. Once the Aborigines realized the nuts were seasonal, they made the most of them at their peak, having discovered that macadamias had a high dietetic and calorific value, thus providing valuable nourishment.

The nuts are very hard to crack, so the Aboriginal tribespeople took them to special areas of flat rock, where they sat and broke open the shells using a smaller rock. These areas can still be seen today and the huge slabs frequently have a bowl-shaped dip in the centre created over the years by being repeatedly thumped to crack open the tough macadamia shells.

Early colonists discovered the macadamia and planted trees – hence the nut became a delicacy. But it took until the 1800s for botanists to discover their potential as a successful food crop. By this time, they had been transported to Hawaii, which subsequently became the world's largest commercial producer. In America, the macadamia is still referred to as the Hawaiian nut. However, in the 1960s, Australia finally started to grow the trees on a commercial basis and by 1995, production levels had overtaken those of America. Now, areas such as New Zealand, California, Israel and South Africa have also started production.

Macadamia nuts have a crunchy texture and a delightful buttery flavour. The oil has an exquisite, slightly sweet flavour that has a natural affinity with tropical fruits, stone fruits and washed rind cheeses. It has a high vitamin and mineral content, oozes with antioxidants, is cholesterol-free and has the highest levels of monounsaturated fats of all the nut and fruit oils, including olive oil.

In most cases, macadamia nuts drop from the trees when they reach maturity. They are then sent to a processor and placed into drying bins. After drying, the nuts are cracked and the kernels separated from the shell; they are then mechanically pressed to extract the oil.

Macadamia oil has a reasonably high smoke point, and works well in stir-fries, but I think its fine, buttery flavour is better suited to salads and salsa, to dress white fish and poultry after cooking and to drizzle over fruits. It works fantastically well in a sort of *pinzimonio* way, i.e. served as a dip for crunchy young veg. Consider it as an alternative when recipes call for a light olive oil.

roasted butternut squash and pancetta salad with pumpkin oil and mixed spice dressing

This combination is a knockout. I use the ground mixed spice usually used for fruit cakes, which is a well-balanced mixture of cassia, ginger, nutmeg, caraway, cloves and coriander – not to be confused with Jamaican allspice, which is the dried berry of the allspice tree and something very different. Ground cinnamon works well, too.

1 large butternut squash, peeled

2 garlic cloves, peeled and crushed

1 teaspoon caster sugar

2 tablespoons pumpkin oil

150 g pancetta lardons

a handful of rocket leaves, to serve

sea salt and freshly ground black pepper

Dressing

6 tablespoons pumpkin oil

2 tablespoons red wine vinegar

1–2 tablespoons runny honey

½ teaspoon ground mixed spice

Serves 4

Preheat the oven to 200°C (400°F) Gas 6.

Deseed the butternut squash and pop the seeds into a sieve. Wash them under cold running water until clean, then pat dry with kitchen paper.

Cut the squash flesh into bite-sized chunks. Put in a roasting tin with the seeds and garlic, sprinkle with the sugar and drizzle with the pumpkin oil. Season and toss well. Roast in the preheated oven for 10 minutes. Remove from the oven, scatter the pancetta lardons on top and return to the oven for a further 15 minutes, or until the squash and garlic are soft and golden and the pancetta is crisp.

To make the dressing, put the pumpkin oil, red wine vinegar, honey and mixed spice in a saucepan, whisk to combine and heat gently over low heat. Season to taste.

Put the roasted squash mixture on four salad plates, drizzle with the warm dressing and top with rocket leaves. Serve immediately.

brioche toasts with roast peaches and sweet macadamia dressing

I often serve this as part of a brunch buffet, but it also makes a lovely dessert. Try it with fresh apricots when they're in season, too.

8 ripe but firm peaches, halved and stoned

3 tablespoons macadamia oil

50 g caster sugar

3 brioche buns, sliced horizontally

Dressing

75 ml macadamia oil

grated zest and freshly squeezed juice of 1 small unwaxed orange

grated zest and freshly squeezed juice of ½ unwaxed lemon

1 tablespoon caster sugar

Serves 4

Preheat the oven to 200°C (400°F) Gas 6.

Put the peaches in a baking dish, cut-side up. Scatter with half the macadamia oil and half the caster sugar and bake in the preheated oven for about 30 minutes, until soft. (Leave the oven on.)

Lightly glaze the brioche slices with the remaining macadamia oil. Scatter with the remaining sugar and put on a baking sheet. Bake in the preheated oven for 5 minutes, or until crisp and golden. Remove from the oven and leave to cool.

To make the dressing, whisk together the macadamia oil, orange and lemon zest and juice, and the caster sugar.

Arrange the brioche toasts on pretty dessert plates and top with the peaches. Drizzle with the dressing and serve immediately.

flavoured oils and marinating in oil

A dash of infused oil is a quick and easy way of transforming even the most humble of ingredients. And oils really are incredibly obliging when it comes to absorbing the flavours of herbs and spices. Add a few chilli peppers to mild-mannered oil and suddenly kick-ass attitude sets in, happily imparting a fiery note to soups and stir-fries. Add a rosemary sprig or two to extra virgin olive oil and try it trickled over fish. A bottle of basil oil will capture the taste of summer and magically carry you through the cooler months. Likewise, ripe summer tomatoes, roasted vegetables, autumn mushrooms, cheeses and all manner of delicious foods can be marinated and preserved in olive oil.

basil oil

I suggest using a fruity Italian oil for this. Some people heat the oil before infusing it, but I prefer not to. Basil grown in the sun will make better oil than the potted supermarket sort.

a large handful of fresh basil leaves

1 litre extra virgin olive oil

Lightly bruise the basil and push into sterilized bottles. Pour in the oil, seal and leave for at least a week before using. Store in a dark place, but not in the fridge.

fennel and fresh lemon oil

You could omit the fennel seeds and just make a lovely lemon-infused olive oil if you prefer.

1 tablespoon fennel seeds

zest of 1 unwaxed lemon, pared with a vegetable peeler

500 ml light and fruity extra virgin olive oil

Toast the fennel seeds in a dry pan until fragrant. Give them a quick crush using a pestle and mortar, or a few hefty thwacks with a rolling pin.
Put them, with the lemon zest, into sterilized bottles. Pour in the oil, seal and leave for at least a week before using. Lovely with salmon.

oven-dried tomatoes in olive oil

Exact quantities of ingredients don't really apply here – it depends how many scrumptious tomatoes you plan to nibble on when they come out of the oven.

- 2 kg midi plum tomatoes, halved
- 3–4 fresh thyme sprigs
- 1–2 teaspoons caster sugar
- sea salt and freshly ground black pepper
- 3–4 tablespoons extra virgin olive oil, plus 100 ml or so to cover

Preheat the oven to its lowest setting. Put the tomatoes on a baking sheet (you'll probably need a couple). Scatter with the thyme and sugar and some seasoning. Bake in the preheated oven for 2–3 hours (depending on the oven temperature), until the tomatoes are dried out but not crisp. Spoon into sterilized jars, cover with olive oil and seal. Store in a cool place, but not in the fridge.

home-made cheese marinated in olive oil and chilli

Choose light, buttery oil that won't overpower the cheese. Try marinating feta or goats' cheese in the same way, too.

- 900 g full-fat Greek-style natural yoghurt
- 1–2 tablespoons fresh thyme leaves, plus 3–4 sprigs
- 3 garlic cloves, peeled and crushed
- about 100 ml light extra virgin olive oil
- 1–2 teaspoons dried chilli flakes
- sea salt and freshly ground black pepper

Put the yoghurt into a large bowl and stir in the thyme leaves and garlic. Season to taste. Line a large sieve with a double layer of muslin or a clean tea towel and sit it over a large bowl. Pour in the yoghurt and leave in the fridge for 24 hours, or until the whey has drained off and the yoghurt is firm. Form the cheese into walnut-sized balls and layer carefully in a clean, sterilized jar, adding the chilli flakes and thyme sprigs. Cover with olive oil and refrigerate.

vinegars

The word 'vinegar' is a combination of the French words *vin* and *aigre* and literally means 'sour wine', and that's exactly how vinegar came about more than 10,000 years ago. An enterprising chap discovered a cask of wine that had oxidized and was undrinkable, and thought to employ it in other ways. By 5000 BC, the Babylonians were using vinegar as a condiment and preservative. At the time of the bubonic plague, it was thought to prevent the spread of germs, and during the First World War it was used to treat wounds.

Vinegar is a solution of acetic acid created through fermentation. In theory, it can be made from most foodstuffs that contain natural sugars. Naturally occurring yeasts convert the sugars into alcohol, then bacteria transforms the alcohol into vinegar. After secondary fermentation, the resulting acetic acid retains the flavours of the original fermented food, hence cider vinegar will taste of apples.

*balsamic vinegar

It seems hard to believe that as little as 30 years ago, balsamic vinegar was virtually unheard of outside Italy and yet balsamic vinegar has been made in the Emilia Romagna region since the eleventh century.

There are two types of balsamic vinegar available to buy, with one fundamental difference between them. Aceto Balsamico di Modena is balsamic vinegar produced in the Modena area, the price and quality of which varies greatly. There are some very good vinegars under this umbrella, but there are also very cheap mass-produced brands on the market too: some are nothing more than white wine vinegar with added caramel.

True balsamic vinegar bears the name Aceto Balsamico Tradizionale di Modena or Aceto Balsamico Tradizionale di Reggio Emilia, and these names are protected by law. This certification verifies that the ingredients and method of production meet the required standard. Production is only permitted within the two provinces, and a panel of expert tasters give the vinegar the mark of approval. It is then sealed with a coloured stamp according to age. Red signifies vinegar that has been aged for up to 50 years, silver a minimum of 50 years, and gold a minimum of 75 years; some are aged for more than a century.

These vinegars are steeped in history and tradition, nurtured with pride throughout the centuries by the wealthy families of Modena and Reggio and passed on through generations as a treasured family heirloom. Making vinegar has long been considered an art form amongst these families. In fact, such was a family's pride in their own vinegar that new barrels were often started to celebrate the birth of a child, to be given years later as the child's wedding gift.

Balsamic vinegar is made from sweet white Trebbiano grapes. The pressings or 'must' from the crushed grapes are boiled slowly to a syrupy concentrate. The strained syrup is then transferred to large oak barrels and the 'vinegar mother' or starter is added. The ageing process begins and over the following years, the vinegar will be moved to a series of smaller barrels, usually on a yearly basis. Producers typically have their own favourite combination of woods and the secret is closely guarded, but most will include juniper, chestnut, ash, cherry and oak. It is this progressive storage in different woods that gives the vinegar its own distinctive character.

If you do splash out on a bottle of expensive *tradizionale* vinegar, save it for dressing and drizzling rather than cooking. Serve it over roasted meats, such as pork and beef, trickle it sparingly over salty cheeses, tender young vegetables and fruits such as strawberries or figs. Try it drizzled over vanilla ice cream, too.

The better-quality vinegars amongst those bearing the name Aceto Balsamico di Modena can be good to use for cooking, but check the label and avoid those that list caramel as an ingredient. Use these in recipes where the vinegar is exposed to high temperatures or bubbled down and reduced.

sticky pork fillet with pecorino, mustard mash and balsamic onions

Tender pork with a golden, rosemary-flecked cheese crust, fluffy mustard-speckled mashed potatoes and sticky, savoury-sweet onions add up to cloud nine dining.

two 400-g pieces of whole pork fillet, trimmed of any surplus fat

3 tablespoons extra virgin olive oil

150 g pecorino cheese, finely grated

a small bunch of fresh rosemary, chopped

sea salt and freshly ground black pepper

Mash

900 g floury potatoes, peeled and cubed

2 tablespoons wholegrain mustard (or to taste)

2 garlic cloves, peeled and crushed

3–4 tablespoons milk

4–5 tablespoons extra virgin olive oil

Onions

2 tablespoons extra virgin olive oil

6 red onions, peeled and sliced

80 ml balsamic vinegar

Serves 4

Preheat the oven to 200°C (400°F) Gas 6.

Brush the pork fillets with 1 tablespoon of the olive oil. Mix the pecorino and rosemary together and spread over a sheet of greaseproof paper. Roll the pork fillets in the mixture, pressing down well so they are evenly coated. Put them in a roasting tin and drizzle with the remaining oil. Roast in the preheated oven for 30 minutes, or until the pork is cooked through and the cheesy crusts are golden. Leave to rest in a warm place.

Meanwhile, to make the mash, boil the potatoes in salted water until soft. Drain and mash. Beat in the mustard, garlic, milk and olive oil. Keep warm.

To make the onions, heat the olive oil in a frying pan and add the onions. Cook for 3–4 minutes, until light golden and starting to soften. Add the balsamic vinegar and 40–50 ml water. Cook slowly for about 10 minutes, or until the onions are soft and sticky, then season to taste.

Slice the pork and serve with the mash and a good dollop of sticky balsamic onions.

balsamic ice cream with crushed strawberries

I remember my youngest son's face when I first asked him to taste ice cream made with balsamic vinegar and black pepper. He looked positively horrified. Now he says he's glad I insisted because it's turned out to be one of his favourites – and mine too. Please choose a good-quality vinegar and make sure the black pepper is coarsely ground: if you're really not keen on the idea of adding the pepper, the ice cream still tastes pretty good without it, but try it once. Like Tim, I think you'll be pleasantly surprised.

100 g caster sugar

3 egg yolks

3 tablespoons balsamic vinegar

300 ml double cream

200 ml whole milk

coarsely ground black pepper, to taste

Crushed strawberries

400 g strawberries

1–2 tablespoons caster sugar (or to taste)

3 tablespoons strawberry liqueur or framboise

an ice-cream maker

Serves 4

Put the sugar, egg yolks and vinegar in a heatproof bowl and beat with an electric beater until thick and creamy.

Pour the cream into a saucepan and heat to simmering point. Remove from the heat and stir into the egg mixture. Return to the pan and stir over very low heat until slightly thickened and custard-like. Remove from the heat, pour into a clean bowl and leave to cool.

When the custard is cold, stir in some coarsely ground black pepper to taste, then transfer to an ice-cream maker and churn until frozen.

About 10 minutes before serving, make the crushed strawberries. Simply mix the strawberries, sugar and framboise together and crush lightly.

Scoop the ice cream into chilled glasses and serve with the crushed strawberries.

*vincotto

Of all the vinegars that I use on a regular basis, vincotto is one of my absolute favourites. Its name literally means 'cooked wine'.

Its production is thought to be based on the old methods used by the people of Carthage to make a sweet wine from semi-dried grapes, or raisins. The wine eventually found its way over to Italy and became very popular during the time of the Roman Empire. Grapes were harvested and left out in the sun before being pressed into wines.

Vincotto is made in a similar way. It is made from two grape varieties, Black Malvasia and Negroamaro. The grapes are either left on the vine to dry, or picked and laid out on wooden frames. They are then pressed, and the resulting 'must' ('must' is the juice of grapes that still contains the skins, seeds, pulp and stems of the fruit) is gently boiled until the mixture has reduced to a mere one-fifth of its original volume, a process which takes around 24 hours. The mixture is then strained and the juice is transferred to aged oak barrels where a 'vinegar mother' or starter is added. The vinegar is left to age for four years.

Vincotto is a sweet, smooth vinegar with a lovely silky texture and a flavour that reminds me of ripe autumn plums and plump semi-dried Agen prunes. It works beautifully in savoury dressings and sauces, but is also fabulous in ice cream recipes or just drizzled over ice cream. I sometimes use it to make a delicious dark chocolate glaze for a dense chocolate and almond cake. It makes a good substitute in recipes where you might use balsamic vinegar, although I'm not suggesting it should replace it in the kitchen storecupboard – rather it's just another string to your culinary bow!

grilled plum and goats' cheese toasties with vincotto

These tasty toasties are easy to rustle up and make a delicious starter or a stylish lunch. You could even make smaller versions to serve as elegant party nibbles.

8 small slices of ciabatta or French bread

3 tablespoons walnut oil

4 ripe plums, stoned and cut into wedges

1–2 tablespoons demerara sugar

120 g goats' cheese, crumbled

2 tablespoons vincotto

freshly ground black pepper

Serves 4

Preheat an overhead grill to high.

Drizzle the bread with the walnut oil and toast on both sides until golden. Arrange the plums on top of the toast, to cover it. Sprinkle with the demerara and grill under the preheated grill until the sugar has melted and is bubbling. Remove from the grill and scatter with the goats' cheese. Return to the grill and grill for a further minute or so.

Drizzle with vincotto and add a good grinding of black pepper. Serve immediately.

pan-fried tuna steaks with warm vincotto-dressed lentils

The combination of earthy lentils and sweet, rich vincotto is stunning. This recipe works well with balsamic vinegar too, but for me, vincotto definitely has the edge.

- 300 g beluga lentils (or Puy if you can't find beluga)
- 3 tablespoons olive oil
- 1 onion
- 100 g pancetta lardons
- 2 tablespoons vincotto
- 4 tuna steaks
- sea salt and freshly ground black pepper
- freshly squeezed juice of ½ lemon

Serves 4

Cook the lentils in boiling water according to the manufacturer's instructions. Meanwhile, heat 2 tablespoons of the olive oil in a frying pan and fry the onion for a few minutes, until softened but not coloured. Add the pancetta and cook until crisp. Drain the lentils and add to the pan. Stir in the vincotto and keep warm.

Brush the tuna with the remaining olive oil and season it. Heat a ridged grill pan or non-stick frying pan and cook the tuna for 1–2 minutes on each side, depending on how thick the steaks are. (Take care not to overcook the fish – overcooked tuna has the texture of old, worn out carpet.) Drizzle a little lemon juice over the fish and serve with the warm lentils.

* sherry vinegar

Sherry is one of the world's oldest wines. The Phoenicians introduced the distinctive Palomino grape to south-west Spain and now, by law, sherry must be produced within a triangular area encompassing Cádiz, Sanlúcar de Barrameda and El Puerto de Santa Maria. Production is largely focused around Jerez and so it has earned the name 'vino de Jerez'.

Sherry vinegar came about as the result of a happy accident. When a cask of a winemaker's sherry was contaminated with acetic acid bacteria, it became undrinkable and turned to vinegar. The vinegar was given to family and friends to use for cooking and kept under wraps – winemakers were embarrassed to admit that some of their wines hadn't made the grade. Eventually, someone decided that sherry vinegar was a very special product in its own right, and worth developing. In 1995, the Jerez DOC (denomination of origin, or mark of quality) became applicable to sherry vinegars. Nowadays, advanced sanitary procedures mean inadvertent contamination is a rare occurrence and vinegar producers must consciously introduce the required species of bacteria into the chosen casks of sherry in order to begin its conversion into vinegar. The added bacteria will transform alcohol and oxygen into acetic acid.

The vinegar is stored in barrels previously used for sherry and left to mature in airy bodegas using the '*solera* and *criadera*' system, a unique system which utilizes rows of 500-litre casks stacked in a pyramid, each cask containing a different aged vinegar. The bottom layer (the *solera*) contains the oldest vinegar. When the vinegar from these casks is ready to be bottled, only one-third is taken. The cask is then replaced with some vinegar from the row above, also called the first *criadera*. The vinegar from that *criadera* is then replenished from the layer above, and so it goes on. The process takes an average of six years, by which time the vinegar has an intense flavour and rich, nutty aroma.

Some vinegars are matured for in excess of 30 years and sell for a fortune, but for everyday use something a little younger will suffice!

Sherry vinegar is a fantastic product, but it has much more of a kick to it than the Italian balsamic vinegars and sweet vincottos, so it should be used sparingly. Take a tip from the Spaniards and add a splash of sherry vinegar to pep up soups, in the way they use it to embellish their gazpacho. Stir it into sauces to serve with grilled fish or poultry – particularly the Spanish-style sauces that are thickened with nuts or bread. Drizzle it over oranges and use it to dress creamy bean salads and dark, leafy vegetables. In dressings, it makes a good partner for the more robustly flavoured oils, such as walnut and pumpkin. If you've never used it before, try it in the stunning Ajo Blanco (see page 78).

mixed griddled fish with romesco sauce

Sherry vinegar is the essential ingredient that lifts this fabulous Spanish sauce. I love it. I always have to make a double quantity because I can't resist spooning some out and dunking a few crisp vegetable sticks in it while I'm cooking...

400 g baby squid, cleaned

4 small pieces of salmon fillet

4 small pieces of cod

4 tablespoons extra virgin olive oil

freshly squeezed juice of 1 lemon

1 garlic clove, peeled and crushed

a small bunch of fresh flat leaf parsley, chopped

sea salt and freshly ground black pepper

Sauce

2 dried sweet peppers

6 plum tomatoes, halved

60 ml extra virgin olive oil

a pinch of caster sugar

2 large slices of good country-style bread, crusts removed

2 garlic cloves, peeled and crushed

50 g whole blanched almonds

2–3 tablespoons sherry vinegar

a pinch of dried chilli flakes, or to taste

Serves 4

Preheat the oven to 180°C (350°F) Gas 4.

To make the sauce, soak the peppers in warm water. Put the tomatoes in a roasting tin, drizzle with 1 tablespoon of the olive oil and sprinkle with the sugar and some seasoning. Roast in the preheated oven for about 20 minutes, or until soft.

Tear the bread into smallish chunks and toss with 2 tablespoons of the olive oil, the garlic and the almonds. Spread out evenly in a roasting tin and roast in the preheated oven for about 10 minutes, or until the bread and almonds are golden brown.

Spoon the roasted tomatoes into a food processor. Drain the soaked peppers and add them to the machine, then whizz until everything is chopped. Add the bread and almond mixture and whizz again until you have a coarsely textured paste. Stir in the sherry vinegar, the chilli flakes and the remaining olive oil. Season to taste and set aside.

Score a crisscross pattern on the squid tubes with a very sharp knife. Brush the squid, salmon and cod with 1 tablespoon of the olive oil each. Heat a griddle pan until hot, then cook the seafood for 2–3 minutes on each side, adding the squid tentacles after a couple of minutes. The fish should be just cooked in the centre and the squid should be golden and tender. Stir the remaining olive oil, the garlic, lemon juice and parsley together and season with salt. Spoon this over the griddled fish and serve immediately with the romesco sauce.

ajo blanco

This soup is a real stunner and yet it's such a doddle to make. If you're in a real hurry, then at a pinch you could save time by using ground almonds, but please don't skimp on the quality of the bread – it's essential to use a good country-style loaf rather than the soggy sliced stuff!

2 thick slices of stale white country-style bread, crusts removed

3 garlic cloves, peeled

150 g blanched almonds, very finely ground

150 ml extra virgin olive oil

2 tablespoons sherry vinegar (or to taste)

a small bunch of white grapes, halved

sea salt and freshly ground black pepper

Serves 4 as a main course

Soak the bread in a little water for 5 minutes or so to soften it. Squeeze out the excess water and transfer the bread to a food processor. Add the garlic, ground almonds and 800 ml water and whizz until smooth. Season to taste. If possible, refrigerate the soup for an hour or so to allow the lovely garlicky flavour to develop.

Stir in 100 ml of the olive oil and the sherry vinegar, adjust the seasoning if necessary and spoon into four chilled bowls. Drizzle the soup with the remaining olive oil and garnish with white grapes. This is lovely served with toasted country-style bread, drizzled with yet more extra virgin olive oil.

*red and white wine vinegars

Apple cider vinegar may be the biggest-selling vinegar around the world (largely due to the fact that it turns up in 'natural' health tonics, potions and pills, as well as being used for culinary purposes) but red and white wine vinegars are still the most popular in the kitchen. It seems the happy accident that occurred around 10,000 years ago, when a batch of wine inadvertently oxidized and turned to vinegar, had quite a knock-on effect!

Both red and white wine vinegars are made from fermented wine. As a general rule, red wine vinegar is made from red grapes and white wine vinegar from white, although some white wine vinegar can be made from red grapes, just to confuse matters.

Red wine vinegar is favoured in many of the Mediterranean countries and is a common storecupboard item in most French households. There are different qualities, and this will depend on a host of factors: the quality of wine used, production, storage, age, etc. Red wine vinegars age well, indeed some red wine vinegar can command prices similar to that of a good wine. Just like wine, as the vinegar matures, the unstable pigments disappear and can leave the vinegar with more of a reddish-brown tinge.

Good red wine vinegar is fun and easy to make at home, but you'll need some vinegar mother (the slimy live starter required to begin the fermentation process). I have my very own vinegar now, thanks to my lovely friends Elaine and Peter, who shared a bit of the 'mum' from their own vinegar. Theirs had rather grand Bordeaux ancestry, whereas my own now has rather a lot of Italian influence, but I must say, it is quite wonderful! Of course, you can buy vinegar starters from many brewers' supply shops. Once you have the starter, any leftover wine you might have can be added to the vinegar, so it's a great way to use up those little bits of wine you're never quite sure what to do with. Just be sure not to use corked wine and keep it in a cool place.

Red wine vinegar is a delicious base for vinaigrettes and dressings and fabulous for deglazing a pan, particularly after cooking red meat and pork. A splash or two will pep up soups and casseroles, but do offer good red wine vinegar at the table, with some extra virgin olive oil, for diners to dress their own salads.

White wine vinegar makes a great base for home-made mayonnaise and is perfect for cutting buttery and creamy sauces such as hollandaise and béarnaise, as well as being a good choice for vinaigrettes and dressings.

easy chilli jam

This recipe couldn't be easier. Whizz everything in a blender, then let it bubble up to a thick, glossy, irresistible conserve that is great with grills, cold meats and cheeses, and is a boon in the kitchen when it comes to adding a little pizzazz to sauces, stir-fries and the like. My fridge feels naked without it.

1 kg ripe tomatoes, quartered
6 red chillies (or to taste)
3 cm fresh ginger, peeled and grated
100 ml white wine vinegar
250 g caster sugar
2 tablespoons extra virgin olive oil

Put the tomatoes and chillies in a food processor and whizz to a purée. Pour into a saucepan with the ginger, vinegar and sugar, and stir over medium heat until the sugar has dissolved. Leave to bubble for a further 30 minutes, or until reduced and thickened.

Add the olive oil and cook for 10–15 minutes, or until the mixture has the consistency of jam. Remove from the heat, leave to cool, then spoon into sterilized jars and refrigerate until needed.

poulet sauté au vinaigre

Typically, this traditional French dish is thickened and enriched with butter. I have used extra virgin olive oil instead, but by all means use butter if you prefer. The one thing that definitely isn't up for negotiation is the use of good-quality red wine vinegar. Cheap vinegar is far too astringent for this dish and will produce a harsh and unpleasant sauce – a far cry from the mouth-watering result you should get.

90 ml extra virgin olive oil

a 2-kg chicken, cut into 8 pieces

500 g very ripe cherry tomatoes (or substitute a 400-g tin cherry tomatoes in juice and rinse before using)

2 garlic cloves, peeled and crushed

200 ml good-quality red wine vinegar

300 ml chicken stock

a small bunch of fresh parsley, chopped

salad leaves, to serve

sea salt and freshly ground black pepper

Serves 4–6

Heat 3 tablespoons of the olive oil in a large frying pan. Season the chicken all over and cook for 3–4 minutes on each side, or until golden. Add the tomatoes and garlic to the pan. Cook for 10–15 minutes, squashing the tomatoes down with the back of a spoon, until they are thick and sticky and have lost all their moisture.

Pour in the red wine vinegar and leave it to bubble for 10–15 minutes, until the liquid has almost evaporated. Pour in the stock, and cook for a further 15 minutes or so, until reduced by half.

Stir in the remaining olive oil and parsley and serve with salad leaves.

*asian vinegar

While there are some fairly potent vinegars made from such ingredients as sugar cane, coconut and palm sugar, which are commonly used in the Philippines and occasionally in Indian cuisine, the best known and most widely available Asian vinegars are the rice vinegars.

As the name suggests, rice vinegars are usually made from fermented rice and originated in Japan and China. Be aware that the terms 'rice vinegar' and 'rice wine vinegar' are used interchangeably on vinegar bottles.

Japanese brown rice vinegars are generally milder and sweeter than the Chinese varieties, and range in appearance from having little or no colour through to a pale straw colour. You may find two types available: rice vinegar and seasoned rice vinegar. The latter is a sort of short-cut sushi vinegar, and comes already seasoned with salt, a little sugar and sometimes sake. Not only is Japanese rice vinegar an essential ingredient in the preparation of sushi rice, but it makes great dressings for savoury salads and some fruit dishes.

White rice vinegar tends to be the most common Chinese rice vinegar, and although it is more acidic than its Japanese counterpart, it's still generally milder than Western vinegars. It is commonly made from fermented glutinous rice. Best for jazzing up stir-fries and salad dressings, it can also be used for pickling – and for sweet dishes too, as you will have discovered in the Sesame and Rice Vinegar Wafers on page 52. Chinese rice vinegar should be widely available in supermarkets and Asian stores, but at a pinch, you could substitute apple cider or white wine vinegar if you have trouble finding it.

Black rice vinegar is usually made with glutinous or sweet rice, although millet or a tropical grass known as 'sorghum' can sometimes be used. Black rice vinegar has a dark, inky colour and a deep, rich flavour that works best in a dipping sauce or in slow-braised meat and duck dishes. I love it stirred into sticky marinades for pork ribs and chicken wings.

With its distinctive red colour, red rice vinegar has an intriguing sweet and sour flavour that comes from the mould found on the unique red yeast rice used in its production. It works well in soups and seafood dishes and is a common ingredient in shark's fin soup. It makes a good dipping sauce, too. Add a little sugar to it and it becomes an acceptable alternative for dishes featuring black rice vinegar.

People often wonder what the difference is between rice vinegar and rice wine, and whether one can be substituted for the other. Chinese Shaoxing rice wine and Japanese sake are both made from fermented rice, but unlike rice vinegars, both can be served as a drink as well as being suitable for use in cooking. Mirin is also a rice wine product, but it is generally used solely for cooking. Recipes that call for rice wine rather than rice vinegar will be aimed at producing different results, so the two should not be interchanged if you are looking to replicate a recipe exactly.

sticky black rice vinegar chicken wings

These sticky little wings make the sort of finger-licking party food that flies off the plate and has everyone looking around for more. I usually make twice as many as I think I'm going to need – on the rare occasions I have leftovers, they're terrific to nibble cold the next day (and they're great tucked into a picnic box, too). The same marinade also works with pork ribs.

12 meaty free-range chicken wings

6 tablespoons black rice vinegar

3 tablespoons Indonesian sweet soy sauce (*ketjap manis*)

2 garlic cloves, peeled and crushed

3 cm fresh ginger, peeled and grated

2 star anise, finely ground

1 tablespoon tomato ketchup

wedges of lime, to serve (optional)

Makes 12 wings

Put the chicken wings in a large bowl. Mix together the black rice vinegar, sweet soy sauce, garlic, ginger, star anise and ketchup and pour it over the chicken. Toss well to ensure all the wings are evenly coated and leave to marinate for about half an hour.

Preheat the oven to 200°C (400°F) Gas 6.

Transfer the wings to a roasting tin, pour any leftover marinade over them and roast in the preheated oven for about 30–35 minutes, turning a couple of times during cooking. The wings should be sticky, cooked through and tender. Serve hot or cold with wedges of lime, if using.

spicy beef brochettes with cucumber and pineapple salsa

Rice vinegar makes such a versatile storecupboard ingredient – its light flavour adds a zing to so many dishes. Here, it adds a kick to the marinade and makes a delicious contribution to the dressing. The salsa works best with fairly big mint leaves that have been very roughly torn – and only right at the last minute.

450 g beef fillet, cut into chunks

2–3 tablespoons muscovado sugar

1 tablespoon soy sauce

4 tablespoons rice vinegar

4 tablespoons extra virgin olive oil

½ teaspoon dried chilli flakes

100 g macadamia nuts, toasted

Salsa

4 cm cucumber, deseeded and diced

½ small pineapple, peeled and diced

a large handful of fresh mint

3 tablespoons rice vinegar

1 teaspoon caster sugar

6 tablespoons extra virgin olive oil

sea salt and freshly ground black pepper

olive oil, for cooking

Serves 4

Put the beef in a large bowl. Mix together the muscovado sugar, soy sauce, rice vinegar, olive oil and chilli flakes and pour it over the beef. Using a mortar and pestle, grind half of the macadamias and add to the mixture. Toss well to ensure all the beef is evenly coated and leave to marinate for at least half an hour.

Meanwhile, to make the salsa, toss the cucumber and pineapple together. Roughly tear the mint leaves and add to the salsa. Chop the remaining macadamias and add them to the salsa. Mix the rice vinegar, sugar and olive oil together and pour it over the salsa. Season to taste.

Heat a non-stick frying pan or griddle pan and brush with a little olive oil. Skewer the marinated beef on to metal skewers and cook on the hot griddle pan for 3–4 minutes on each side. Serve with the salsa.

*apple cider vinegar

Apple cider vinegar is one of the oldest and most well-known vinegars – its origins go way back to the time of the diligent Babylonians. Thousands of years on, it is currently the most popular of all the vinegars used worldwide.

Renowned for its health-giving properties, not only does apple cider vinegar come in bottles for culinary use, but it finds its way into all manner of pills, potions and tonics. The old saying 'An apple a day keeps the doctor away' means it's hardly surprising that a preparation made from the pulp of crushed apples enjoys a similar reputation.

The base alcohol for the vinegar is cider. In general production, sugar and yeast are added to apple juice and the fermentation process begins. A vinegar mother is then added to the cider and the secondary fermentation process converts it into vinegar. Some artisan vinegars are made using the 'lees' (the settlement) from the cider rather than the cider itself. There are also producers who don't add sugar to the apple juice; this gives the cider base a lower alcohol level and the result is a vinegar that is less acidic.

Cider vinegar contains a whole host of important vitamins, minerals, and trace elements. It is thought to help lower LDL (bad) cholesterol, and help with conditions such as water retention, high blood pressure, circulatory problems, arthritis and much more. Surprisingly for something that is a vinegar, it has a lower PH level than that of most commercially produced cola drinks!

It tastes fabulous, makes great dressings and mayonnaise, and adds flavour to sauces, marinades, chutneys and pickles. It also makes a great base for home-made flavoured vinegars, too.

mushrooms marinated with raisins and apple cider vinegar

Mushroom fans will love this tasty mix of juicy mushrooms in a sweet and sour chilli-spiked marinade. It makes a perfect side dish for barbecues. Add a crumbling of salty cheese and you have a pretty special main course for vegetarians, too.

60 ml extra virgin olive oil

2 shallots, peeled and finely chopped

2 garlic cloves, peeled and crushed

500 g brown mushrooms, halved

20 ml apple cider vinegar

2 tablespoons raisins

2 tablespoons runny honey

a pinch of dried chilli flakes

fresh oregano leaves, to garnish (optional)

Serves 4–6

Heat 3 tablespoons of the olive oil in a frying pan and fry the shallots and garlic over low heat for 2–3 minutes, until softened. Add the mushrooms and sauté for 4–5 minutes, until golden. Add the cider vinegar and raisins, and bubble for a minute or so. Stir in the honey, remaining olive oil and chilli flakes. Cook for a further minute. Remove from the heat and leave to cool. Leave to marinate for half an hour before serving. Garnish with oregano, if using.

flavoured vinegars

Creating your own flavoured vinegars at home means that you can make the most of seasonal fruits, herbs and vegetables and enjoy them throughout the year. Start with a good-quality wine or cider vinegar as a base, rather than one of the commercial white distilled vinegars.

raspberry and thyme vinegar

You should find funnels at good kitchenware shops or home brewing suppliers.

1 kg raspberries

1 litre white wine vinegar
or apple cider vinegar

2–3 sprigs of fresh thyme

Makes about 1 litre

Put the raspberries and white wine in a non-reactive bowl. Crush the raspberries with a wooden spoon to release their juices. Transfer the mixture to a large sterilized container and seal. Leave for 3–4 weeks, shaking at regular intervals.

Line a funnel with a double layer of muslin or a clean tea towel. Strain the vinegar into clean, sterilized bottles. Pop in the thyme, seal the bottles and label with the date. The vinegar is now ready to use. Lovely in salads and sauces to go with chicken, mild cheeses and pâtés.

blackberry vinegar

Be sure to remove the stalks and discard any fruit that has mould on it before checking that you have the weight stipulated!

500 g blackberries

1 litre white wine vinegar or apple cider vinegar

Makes about 1 litre

Put the blackberries and white wine in a non-reactive bowl. Crush the blackberries with a wooden spoon to release their juices. Transfer the mixture to a large sterilized container and seal. Leave for 3–4 weeks, shaking at regular intervals.

Line a funnel with a double layer of muslin or a clean tea towel. Strain the vinegar into clean, sterilized bottles. You may like to add two or three blackberries for presentation. Seal the bottles and label with the date. Use for dressings, marinades and sauces – it goes especially well with duck and venison.

mango vinegar

This pretty, orange-coloured vinegar looks as lovely as it tastes. If you're a fan of spiciness, add a (clean) hot bonnet chilli to the bottles!

3 very ripe mangoes

1 litre white wine vinegar

2 tablespoons caster sugar

Makes about 1 litre

Remove the stones from the mangoes and put the flesh in a food processor. Whizz to a purée. Pour the vinegar into a saucepan and add the sugar. Bring to the boil. When the sugar has dissolved, add the mango pulp and simmer for 3–4 minutes. Leave to cool. Line a funnel with a double layer of muslin or a clean tea towel and sieve the vinegar into sterilized bottles. Seal the bottles and label with the date. The vinegar can be used straightaway. Great to splash into Caribbean-style sauces and dressings to go with pork, venison and duck.

rosemary and lemon vinegar

I love this vinegar. Try to use an unwaxed lemon if possible.

1 unwaxed lemon

2 sprigs of fresh rosemary

1 litre white wine vinegar or apple cider vinegar

Makes 1 litre

Peel the zest from the lemon using a vegetable peeler (be careful to discard all the white pith). Put it into a bowl with the rosemary. Heat the white wine to boiling point and leave to cool a little. Pour into sterilized bottles, add the zest and rosemary, seal and label with the date. Leave for at least a week before using. Great in dressings to accompany chicken and fish.

suppliers

Delicioso
www.delicioso.co.uk
Unit 14
Tower Business Park
Berinsfield
Oxon OX10 7LN
Tel: 01865 340055
An online Spanish delicatessen which sources all its products directly from Spain and endeavours to use only family-run enterprises that use traditional methods whenever possible. Stocks lots of lovely olive oils and vinegars, including flavoured varieties of both.

The Gift of Oil
www.thegiftofoil.co.uk
The Enterprise Centre
Washington Street
Bolton BL3 5EY
Tel: 01204 559555
A great range of extra virgin olive oils, including infused types, as well as an excellent supply of quality balsamic vinegars, some aged up to 100 years! Also have flavoured balsamic glazes and attractive oil and vinegar pourers.

The Olive Oil Club
www.oliveoilclubs.com
3 Waters Edge
Ellworthy Park
Frome
Somerset BA11 5LT
Tel: 01373 471836
The best olive oil and balsamic vinegar sourced and tasted by true connoisseurs. Membership offered for olive oil enthusiasts.

Select Spain Gourmet
www.selectspain.co.uk
298 Wellingborough Road
Northampton NN1 4EP
Tel: 0845 644 0774
Stockists of Spanish gourmet foods, including extra virgin olive oil and gourmet vinegars.

Merchant Gourmet
www.merchant-gourmet.com
Tel: 0207 635 4096
Good range of quality fruit, nut and seed oils, as well and lots of vinegars.

Francesco Accurso
Francescoaccurso@italianmarket.info
70 Minet Avenue
London NW10 8AP
Francesco Accurso runs Italian markets which go around the UK, and he stocks fabulous oils and vinegars from small producers. Email him to find out where the next market is taking place.

Somerset Cider Vinegar Company
www.somersetcidervinegarco.co.uk
5 Centre Road
Edington nr. Bridgwater
TA7 9JR
Award-winning makers of artisanal apple cider vinegar, endorsed by chefs Mark Hix and Rick Stein.

Daylesford Organic
www.daylesfordorganic.com
Suppliers of organic vegetables, cheese, pastries, meat and other stock such as flavoured vinegars.

index

aïoli, garlic and walnut, 43
ajo blanco, 78
apple cider vinegar, 90
argan oil, 26
  lamb and butternut squash tagine, 28
  spiced argan, lime and honey dressing, 9
Asian vinegars, 84
avocado oil, 30–1
  avocado oil, lemon and pistachio cake, 34
  Italian vegetable and bread soup, 33

balsamic vinegar, 64–5
  balsamic ice cream, 68
  balsamic onions, 67
basil oil, 60
beef brochettes, 88
blackberry vinegar, 93
bread: black grape schiacciata, 21
  fig and hazelnut breakfast bread, 46
  Italian vegetable and bread soup, 33
brioche toasts with roast peaches, 58

cakes, 25, 34
cheese: grilled plum and goats' cheese toasties, 71
  home-made cheese marinated in olive oil and chilli, 61
chicken: a bit like bang-bang chicken, 51
  marinated chicken, raisin and chilli salad, 44
  poulet sauté au vinaigre, 83
  sticky black vinegar chicken wings, 87
cider vinegar, 90
  mushrooms marinated with raisins and, 91
coconut dip, 52
cucumber and pineapple salsa, 88

fennel and chilli dressing, 9
fennel and fresh lemon oil, 60
fig and hazelnut breakfast bread, 46
fish with romesco sauce, 77
flavoured vinegars, 92–3

garlic and walnut aïoli, 43

hazelnut oil, 39
  fig and hazelnut breakfast bread, 46
  hazelnut and orange dressing, 9
  hazelnut dressing, 44

ice cream, balsamic, 68
Italian vegetable and bread soup, 33

lamb: griddled lamb fillet, 43
  lamb and butternut squash tagine, 28
lemon: avocado oil, lemon and pistachio cake, 34
  fennel and fresh lemon oil, 60
  lemon, honey and thyme dressing, 9
  rosemary and lemon vinegar, 93
  sweet lemon and olive oil conserve, 46
lentils, warm vincotto-dressed, 73

macadamia oil, 55
  sweet macadamia dressing, 58
mango vinegar: mango, hazelnut and ginger dressing, 9
marinated mushrooms, 91

olive oil, 11, 13–17
  ajo blanco, 78
  black grape schiacciata, 21
  home-made cheese marinated in, 61
  hot garlic prawns, 18
  olive oil and orange cake, 25
  oven-dried tomatoes in, 61
  poulet sauté au vinaigre, 83
  rosemary and roast potato tart, 22
  sweet lemon and olive oil conserve, 46

peaches, brioche toasts with, 58
pesto, warm walnut, 40
plum and goats' cheese toasties, 71
pork fillet, sticky, 67
potatoes: mustard mash, 67
  rosemary and roast potato tart, 22
poulet sauté au vinaigre, 83
prawns, hot garlic, 18
pumpkin oil, 54
  pumpkin oil and mixed spice dressing, 57

raspberry and thyme vinegar, 92
ravioli, roast onion and celeriac, 40
red wine vinegar, 80
  poulet sauté au vinaigre, 83
  red wine vinaigrette, 9
rice vinegars, 84
  sesame and rice vinegar wafers, 52
  spicy beef brochettes, 88
  sticky black vinegar chicken wings, 87
romesco sauce, 77
rosemary and lemon vinegar, 93
rosemary and roast potato tart, 22

salad dressings, 8–9
salads: a bit like bang-bang chicken, 51
  marinated chicken, raisin and chilli, 44
  roasted butternut squash and pancetta, 57
salsa, cucumber and pineapple, 88
schiacciata, black grape, 21
sesame oil, 48–9
  a bit like bang-bang chicken, 51
  sesame and rice vinegar wafers, 52
  toasted sesame, lemongrass and ginger dressing, 9
sherry vinegar, 74–5
  ajo blanco, 78
  mixed griddled fish with romesco sauce, 77
soups: ajo blanco, 78
  Italian vegetable and bread soup, 33
spinach, hot garlic prawns with, 18
storing oils, 14
strawberries, balsamic ice cream with, 68

tomatoes: easy chilli jam, 81
  oven-dried tomatoes in olive oil, 61
  poulet sauté au vinaigre, 83
tuna steaks with warm vincotto-dressed lentils, 73

vinaigrettes, 8–9
vincotto, 70
  grilled plum and goats' cheese toasties with, 71
  pan-fried tuna steaks with warm vincotto-dressed lentils, 73

wafers, sesame and rice vinegar, 52
walnut oil, 38
  garlic and walnut aïoli, 43
  walnut and blackberry vinegar dressing, 9
  warm walnut pesto, 40
white wine vinegar, 80
  blackberry vinegar, 93
  easy chilli jam, 81
  mango and chilli vinegar, 93
  rosemary and lemon vinegar, 93
wine vinegars, 80